Of Molecules and Men

Of
Molecules
and
Men

By
FRANCIS CRICK

University of Washington Press
Seattle and London

Of Molecules and Men is a volume in the
John Danz Lecture Series

The John Danz Lectures

IN October, 1961, Mr. John Danz, a Seattle pioneer, and his wife, Jessie Danz, made a substantial gift to the University of Washington to establish a perpetual fund to provide income to be used to bring to the University of Washington each year ". . . distinguished scholars of national and international reputation who have concerned themselves with the impact of science and philosophy on man's perception of a rational universe." The fund established by Mr. and Mrs. Danz is now known as the John Danz Fund, and the scholars brought to the University under its provisions are known as John Danz Lecturers or Professors.

Mr. Danz wisely left to the Board of Regents of the University of Washington the identification of the special fields in science, philosophy, and other disciplines in which lectureships may be established. His major concern and interest was that the fund would enable the University of Washington to bring to the campus some of the truly great scholars and thinkers of the world.

Mr. Danz authorized the Regents to expend a portion of the income from the fund to purchase special collections of books, documents, and other scholarly materials needed to reinforce the effectiveness of the extraordinary lectureships and professorships. The terms of the gift also provided for the publication and dissemination, when this seems appropriate, of the lectures given by the John Danz Lecturers.

Through this book, therefore, the fifth John Danz Lecturer speaks to the people and scholars of the world, as he has spoken to his audiences at the University of Washington and in the Pacific Northwest community.

"Exact knowledge is the enemy of vitalism."

Preface

THIS IS not a scholarly work on the history of vitalism. If it were I should have had to refer at some length to earlier writers like Hans Driesch and Henri Bergson, to say nothing of contemporaries such as Teilhard de Chardin and Michael Polanyi. I am very much out of sympathy with both these thinkers, fashionable though they may be in literary circles. Polanyi, for example, believes that a steam engine cannot be totally described in terms of physics and chemistry. It is clear that he and I are using this expression in quite different ways. According to his terminology an enzyme molecule (which can be thought of as a machine at the molecular level) could not be described in terms of physics and chemistry, a position that seems to me to be ridiculous. As he does not really believe in the central importance of natural selection, it is not surprising that he runs into logical and philosophical difficulties of all kinds.

Rather I have taken vitalism as a theme, and have written down the sort of thing that I find my friends and colleagues are saying about our present and future knowledge of biology. It is thus not a literary or philosophical work, but a collection of scientific results and ideas. Most of the ideas I mention are now fairly commonplace among scientists engaged in extending the frontiers of biological knowledge. It is my hope that their point of view may be of interest to a wider circle.

A few years ago I gave a lecture to the Cambridge Humanists on this topic. But for the invitation to deliver the John Danz Lectures I doubt if this present book, which is a much expanded version of the earlier lecture, would have been written. I thank the Board of Regents of the University of Washington for the opportunity to enlarge and record my thoughts in this way.

The lectures were delivered at the University of Washington in February and March, 1966, under the title, "Is Vitalism Dead?" The statement by Salvador Dali at the opening of the first lecture is quoted with permission from *Playboy Magazine*'s interview with Dali as it appeared in the July,

1964, issue. The quotation from Michael McClure (second lecture) is from "Peyote Poem," published as *Semina 3* by Wallace Berman, 1957. The passage by Thomas Hunt Morgan (third lecture) is quoted from *The Physical Basis of Heredity* (Philadelphia and London: J. B. Lippincott Co., 1919).

organisms as the bacterial cell. Studying the cell as a living unit enables us to see in a broad way what it does. The system is usually too complicated to enable one to deduce from such experiments the *exact* details of the mechanism inside the cell. Breaking open the cell and studying the bits enables one to find the exact behavior, but the breaking process may produce artifacts. To avoid these one must go back to the study of the intact cell. In short it would be difficult to deduce the detailed functions of a watch either from the intact, unopened running watch, or from the smashed pieces, but a combination of these two approaches would tell us most things about the mechanism.

This argument—that one should study both the whole and its parts, the relative emphasis in any particular case being a matter of tactics—applies to all levels in biology. It is an old idea that a biological system can be regarded as a hierarchy of levels of organization, the "wholes" of one level being the parts of the next. Thus cells are the "wholes" of cellular biology, but the parts of tissue biology, and so on. In my view a simultaneous attack at more than one level will, in the long run, pay off better than an attack at a single level, even though in the short run one may concentrate on one level at a time.

Thus eventually one may hope to have the whole of biology "explained" in terms of the level below it, and so on right down to the atomic level. And it is the realization that our knowledge on the atomic level is secure which has led to the great influx of physicists and chemists into biology. Incidentally the above remarks about "wholes" do not apply exclusively to biology. They are also true, for example, of chemistry. A benzene molecule "is more than the sum" of six carbon atoms and six hydrogen atoms. Here the semantic point to watch is the word "sum." Benzene is certainly not an arithmetical sum of its component parts, but nevertheless a theoretical chemist can deduce its behavior by doing the sum the right way; that is, by using the methods of quantum mechanics.

At this point we need to clear up a point about biological explanation. This is always of two types. In examining every biological system one can always ask how it works; meaning how, from a knowledge of its parts, one can predict its behavior. Alternatively one can ask how the system got that way; in other words, how it evolved.

Now these two explanations are in fact very different, for this reason. Most biological organisms, as it happens, work rather reliably. Moreover many of them can be obtained in almost identical copies. Hence one can usually make "predictable" observations on them, at least as well as on many purely physical systems, such as, for example, the eddies in a stream.

On the other hand it is unlikely that the *evolution* of one species into another has this character. This depends upon rare events (such as mutations) and what may be chance factors in the environment. We have no idea of what actually permitted the mammals to evolve from reptiles, and it may not be easy to find out, even though we can see that, having evolved, they are in many ways superior to ("fitter" than) reptiles.

Consequently there is real doubt whether the actual process of evolution is predictable. It may be history rather than science. That is, chance may produce effects which basically alter the historical process. I think this distinction—between the behavior of an organism and its evolution—is of crucial importance, even if some borderline cases cause difficulty. I believe that it is this which is confusing Elsasser and Polanyi in their approach to biology.

I can hardly delay any longer considering what I mean by the word "vitalism." As before, an exact definition is a difficult one, but in a general way it is very easy to see what it means. It implies that there is some special force directing the growth or the behavior of living systems which cannot be understood by our ordinary notions of physics and chemistry. The difficulty in the definition is in explaining what sort of thing this might be, and almost everyone who has written or spoken about it becomes very confused at this point.

It will be my general thesis that the reason the need is felt for a doctrine like vitalism is because we see a complicated pattern of behavior which we cannot easily explain in terms of the concepts that are immediately available to us. It thus is useful to examine in a little more detail those areas of biology where these difficulties are most apparent. It seems to me that there are three such areas. The first one, and the one about which I shall have most to say, concerns the borderline between the living and the nonliving. Not so many years ago this seemed infinitely mysterious, and only by a very considerable act of faith could one believe that an explanation would be possible in terms of physics and chemistry. The second area, which is actually rather closely related to the first one, concerns the origin of life, an event that took place a very long time ago and about which at the moment it is not easy to say very much of scientific value. The third area that causes us to be uncomfortable when we think about it in exact scientific terms is that embraced by the word "consciousness." It deals with the exact behavior of the brain and the explanation in terms of that behavior of our subjective feelings and emotions. This again is an area about which, at the moment, we know rather little.

It should be noticed that in the past there were at least two other sensitive areas about which people worried. Originally it was thought that there was something radically different about organic molecules which made them impossible to synthesize by ordinary chemical methods. This particular worry, which came at a rather early stage in the history of chemistry, was dramatically upset when Friedrich Wöhler succeeded in synthesizing urea, an undoubtedly organic molecule, starting from ammonium cyanate, which was considered to be an "inorganic," or at least nonliving,

17

chemical. In fact Wöhler made it from dried blood, hoofs, and horns. A more definitive proof of the synthesis of various organic compounds from their chemical elements was given by Marcelin Berthelot in 1860. The old belief is, of course, the origin of our phrase "organic chemistry."

The second region which it was thought was going to be difficult to explain concerned the energy produced by living things. There was for a period a real doubt as to whether the elementary laws of thermodynamics were obeyed by living organisms. However it has been shown that there is nothing in the metabolism of organisms which appears to contradict these laws. Both these issues now seem to be thoroughly dead.

All the three areas I have mentioned are brought into mind when we use the ambiguous word "alive." I have wanted to argue that when I say "alive" I mean the difference between the living and the nonliving, but for many people the word "alive" is used in a different sense—one says of somebody, "But he's so *alive!*" Used in this way the term usually refers to the behavior of an animal. I doubt if one would make this statement about a plant. A person is usually said to be "alive" because of the nature of his responses. I do not think one would say of a sleepwalker that he or she was "so alive" in this particular sense.

The reader should not think that vitalistic ideas belong only to the past and that more recent writers have avoided them. On the contrary, in the last few years there seems to have been a resurgence of such ideas. Let me give three examples. Dr. Walter Elsasser, the distinguished physicist, has gone so far as to write a whole book on the subject, entitled *The Physical Foundation of Biology*.* I shall not comment on his remarks about the nervous system, but confine myself to his ideas about reproduction and embryology. Although the book was published in 1958, and although it has a few passing references to the recent developments in molecular biology, it was clearly conceived in an earlier era, and it is a beautiful example of the confusion that can be brought about by ignorance. For example, Dr. Elsasser is surprised to find that genetic information stored in a lobster is not in its hard external parts, but in its "soft" tissue, as he calls it. He seems to think that information so stored will be very easily disturbed by thermal noise. Now we know, as I shall explain in the next chapter, that it is in fact stored on a polymer, the

*London: Pergamon Press, 1958.

chemical bonds of which are almost immune to purely thermal damage. We also know that the cell has various repair mechanisms to correct at least some of the little damage that does occur. Thus the simple facts of chemistry, which as a physicist he appears to ignore, allow the cell to store an immense amount of information, very stably, at the molecular level.

This leads to my second point. His whole argument is based on the apparent dilemma that the large amount of information needed to construct the adult organism cannot conceivably be stored in the germ cells. It cannot be stated too strongly that his dilemma is false. Elsasser has apparently not realized how much information can be stored on the chromosome—it may be as high as 10^{10} bits—and he has probably vastly overestimated the information actually needed to build a human being. Very much of our anatomy and physiology is repetitive, or partly repetitive, or due to the combination of various influences. Let me admit that we do not know at the present time how much information is needed to make a hand, with all the bones, joints, muscles, blood vessels, and so forth in the right places. But we can conceive ways that it could be done which are well within the information capacity available to us. Thus the mechanist's position that such information is stored in the germ cells is still tenable until detailed studies have shown it to be wrong. It may take us many years to attain enough knowledge even to try to construct a dilemma of the type Elsasser has suggested in his light-hearted way.

To get out of his difficulty Elsasser is compelled to invent what he calls "biotonic phenomena." These are "phenomena in the organism that cannot be explained in terms of mechanistic function." He postulates that they are guided by biotonic laws which, he claims, are compatible with the ordinary laws of physics, but additional to them. He has tried to point out how this might arise because in a biological system there may be too many possibilities to be averaged over in the time available. This may turn out to be true in certain contexts, but I fail to see that it will necessarily lead to new laws. In my opinion it is more likely to lead to chance effects. However, as his dilemma is not a real one his explanations need not detain us.

Dr. Elsasser, I suspect, would not like to be classed as a vitalist although he concedes that "we seem to have moved rather close to a vitalistic philosophy." It would perhaps be more correct to call him a "neovitalist": that is, a man who believes in vitalistic ideas while denying that he does so! His arguments would perhaps have seemed convincing thirty years ago, but today no molecular biologist takes them seriously. The detailed knowledge we have acquired in recent years, part of which is outlined in the next chapter, has made it possible to see how misleading general arguments can be when they are not securely based on intimate scientific knowledge. When facts come in the door, vitalism flies out of the window.

My second example is a long article in *Nature* (199:212-19) written by Dr. Peter Mora in 1963 with the provocative title, "Urge and Molecular Biology." A few quotations will make clear the general content of this article. He begins, "Living entities, at all levels in almost all their manifestations, have something of a directed, relentless, acquiring and selfish nature, a perseverance to maintain their own being and a continuous urge to dominate their surroundings, to take advantage of all possible circumstances, and to adjust to new conditions." This shows the meaning with which he is using the word "urge." He goes on to say, "It declines with age and ceases with death, but its existence, while it lasts, makes living entities recognisable and distinct from non-living." And then he asks, "Are present molecular approaches sufficient to lead us to the understanding of the difference between the living and the non-living, or more explicitly, of what I call the 'biological urge'?" The article itself contains much detailed knowledge of molecular biology, all of which is scrupulously correct, but still leaves one with a confused feeling. In fact, he says himself, "I am afraid I cannot be very definite, because I cannot see clearly the nature of the new method, or the extension of the physical methods, I feel it is desirable to find in biology."

A third recent example comes from Dr. Eugene Wigner, the distinguished physicist and Nobel laureate. Wigner has written a paper trying to show, by the methods of quantum and statistical mechanics, that it is impossible to have a self-reproducing system.* He is clearly uncomfortable about this conclusion, and he makes a certain number of reservations that allow him to escape from this dilemma. I do not wish to discuss here the

---

*In *The Logic of Personal Knowledge: Essays Presented to Michael Polanyi on His Seventieth Birthday 11th March 1961,* edited by Edward Shils (Glencoe, Ill.: Free Press, 1961).

validity of this argument. I want to draw attention to the beginning of his final paragraph, which reads, "The writer does not wish to close this article without admitting that his firm conviction of the existence of biotonic laws stems from the overwhelming phenomenon of consciousness." The point to notice here is that because he is worried about consciousness he has written a paper about reproduction. Because he feels there is something wrong in one of our delicate areas, the area of the brain, he has considered another of our delicate areas, the region between the living and the nonliving. It would be difficult to find a more striking example of the type of discomfort that appears to motivate these vitalistic ideas.

In all fairness I think I must say that I think this association also works in reverse. That is, I believe the motivation of many of the people who have entered molecular biology from physics and chemistry has been their desire to *disprove* vitalism. It is interesting to note that, since we now feel that the mystery has been removed at the borderline between the living and the nonliving, a number of molecular biologists are considering moving into the nervous system. Let me draw your attention to a piece of documentary evidence along

24

these lines. This comes in the book called *The Molecular Biology of Bacterial Viruses* by my friend Gunther Stent.* He has a short postscript at the end in which he discusses the question of vitalism. His concluding paragraph starts with the question, "What fundamental questions remain after the solution of the problem of self-reproduction to which the molecular biologist might now address himself?" and he goes on, at the end of the paragraph: "But one major function of biological inquiry would still remain even after origin of life and morphogenesis have also been satisfactorily accounted for—the higher nervous system. Its fantastic attributes now seem to pose as hopelessly difficult and intractably complex a problem as did the phenomena of heredity half a century ago. Perhaps just because of the challenge constituted by the present impossibility of even imagining any reasonable molecular explanation for such manifestations of life as consciousness and memory, the neuron bids fair to become the phage of the future."

Indeed, while on this subject of motivation, it would be of interest for somebody to inquire into the religious faith of these various writers. I have

*San Francisco, Calif.: W. H. Freeman and Co., 1963.

a strong suspicion that it is the Christians, and the Catholics in particular, who write as vitalists, and it is the agnostics and atheists who are the anti-vitalists. Whether this is widely true I must confess I do not know.

Let us return now to a consideration of the meaning of vitalism. We have seen that it means that there must be something else in a biological system which cannot be included under the heading of physics or chemistry. This must be some sort of force, or some directing spirit—this is the sort of idea that appeals to nonscientists. Scientists on the other hand often prefer to think that there will be extra laws in biological systems which are not included in physics and chemistry. The difficulty with this last point is to say clearly what sort of thing such an extra law would be, and, even more, to give a concrete example of one. In a certain sense it could be argued that natural selection is such an extra law, and I would certainly think it was a law of the most fundamental importance for biological systems. But it is not at all clear that something like it could not have been deduced from the study of, say, the chemistry of open systems. It is quite conceivable that the study of chemical engineering would have thrown up a

26

law similar to natural selection, since if we came across anything in the form of a replication process in a simple chemical system we would be confronted with behavior of this kind. Such behavior, however, may be exceedingly rare.

But what test could be used to refute vitalism? One obvious approach would be to try and make some biological object completely synthetically, starting from chemical elements. It would certainly disturb many people if some simple living organism could be made in this way. Nevertheless I foresee that even if it were done there would be those who would claim that the system, having been made by us, was then "colonized" by the vitalistic force, or the soul, or whatever they might like to call it, which thus took over the workings of the system. So in addition to being able to synthesize it we see that it is necessary to explain the behavior of the organism in terms of physical and chemical laws. We must, however, be careful about trying to explain too much. It may well be, as I have already suggested, that the whole course of evolution cannot be easily explained in the way one would explain a physical system—there may have been chance factors, which at certain stages deflected it. Had they been otherwise, quite different

27

higher animals and plants might have resulted. Evolution may, therefore, for all we know, not be so susceptible to exact prediction, but of course this is true of purely physical systems as well. For example, the precise shape of the clouds in the sky at noon tomorrow is not something that we aim to predict in the near future.

In the following chapters I shall set out what we know about our three sensitive areas: that is, the area of molecular biology, of the origin of life, and of the higher nervous system. Finally I have considered the position of vitalism today.

The Simplest Living Things

"THIS IS THE POWERFUL KNOWLEDGE
We smile with it."

MICHAEL MC CLURE

IN THIS CHAPTER I AM GOING TO EXAMINE IN SOME detail what we know about the borderline between the living and the nonliving. At first sight it might appear sensible to start off by looking at virus structure and virus behavior, but for reasons that will become apparent I shall leave them till later. Instead I propose to consider a typical micro-organism, one which can grow in a relatively simple and well-defined chemical medium, as this sort of cell is one of the simplest complete living systems that we can find. The bacterial cell I shall be describing will usually be *Eschericia coli,* although I may from time to time make remarks about other organisms.

Such a cell can grow either in a simple medium or in a complex one. The simple medium consists of, for example, a sugar as a source of both carbon and energy, perhaps ammonia as a source of nitrogen, and phosphate, sodium, potassium, magnesium, and so forth. There is nothing in the medium that we could not synthesize from the elements if we so wished. In a richer medium containing many of the small organic compounds, which otherwise the cell would have to make for itself, it can grow much faster, and under these conditions the cell may divide every twenty minutes. After ten hours, provided there is enough "broth" (as this richer medium is often called), there will be about one thousand million cells, all descendants of the first cell used to start the bacterial culture. To get this rate of growth it is necessary to keep the medium at the right temperature, in this case about blood temperature, and bubble air through it. However oxygen is not essential for all forms of life, and some organisms in fact can grow only in the total absence of oxygen. Oxygen is essential for higher organisms, like ourselves, but many lower organisms can do without it, although if it can be used it always makes the utilization of the food supply more efficient.

We can examine such a cell in the microscope. It is in fact rather small, perhaps one or two microns in diameter (a micron is one thousandth of a millimeter). It is visible in the light microscope, but it is not easy to see any details of the interior of the cell. This can be done, however, by looking with the electron microscope at thin sections of cells which have been fixed and stained, and by this technique one can see a certain amount of structure within the cell. Most of our knowledge has not come in this way, however, but by an intricate combination of many different biochemical and genetical experiments which have told us in considerable detail what goes on inside a cell of this type.

The cell, when we come to look at it, is essentially a bag having a stiff outer wall to give it strength, and just inside that a special filter—a membrane— which effectively divides the inside from the outside. That is, it is impermeable to many of the organic molecules that are produced inside the cell, so that they are retained inside, but it allows other molecules to flow in and out. In addition, by mechanisms which we do not yet fully understand, it is the seat of various metabolic pumps that concentrate inside the cell molecules from the medium, so that they come to a higher concentration where they are needed.

If we look inside this bag we are struck by the fact that there are several areas that appear to

33

contain a very long, thin molecule, looking rather like an enormous piece of string, though naturally on a much smaller scale. I shall have more to say about this presently.

Inside the rest of the bag we find a good number of very large molecules and many rather small ones. To give you a rough idea of the number of different types of molecules, there may be perhaps several thousand different kinds of large ones and perhaps a similar number of small ones. The cell therefore is an intricate chemical factory, in which many chemical reactions are going on simultaneously, small molecules being changed one into another, and large molecules being built out of small ones.

It is one of the interesting generalizations of biochemistry that relatively few medium-size molecules are made by living cells. By this I mean molecules of molecular weight of a few thousand. The reason for this is that in many cases it is more difficult to make a large molecule than a small one, since if it is at all complicated many more chemical bonds have to be made. There are special methods for making large polymer-type molecules, by joining similar small molecules end to end, but such mechanisms tend to produce molecules of considerable length. Thus on this simple picture the molecules in cells are either large polymers, or the small molecules used to form these polymers, or relations of the latter. It is not worth the cell's while to build up complicated molecules except by simple repetitive processes. Thus medium-size molecules are rarely produced.

What is remarkable about the cell is that from very simple components it can build up this enormous variety of molecules of different types and of different sizes. Clearly there has to be some general method by which all these chemical operations can be carried out. The principle is a relatively simple one and has been known now for many years. For every particular chemical reaction—every simple chemical step—there is a special catalyst that speeds up that step and that step only. Whereas at room temperatures such a reaction will take place spontaneously only exceedingly slowly if at all, under the influence of a catalyst the reaction can go very fast and become an important one in the life of the cell. These catalysts are known as enzymes, and it is one of the most important generalizations of biochemistry that, although each type of enzyme is different from every other type, all enzyme molecules belong chemically to the same family. Although they are all different they are all built in the same sort of way, on the same general plan, and are therefore classed together. They are in fact the protein molecules of the cell, and, although protein can be used for other purposes (as structural material, for example), its main and most important function is to provide the enzymes—the specific catalysts—of the living cell.

This little bacterial cell then is, in a very small space, a highly efficient and very intricate chemical factory. Into it flow the raw materials from the medium. Inside it the protein molecules catalyze the changing of these raw materials into other small organic molecules of a great variety of types, and other proteins then act to join these together into the macromolecules that make up the cell. All these metabolic processes require energy, which comes from the breaking down of the raw material provided by the medium; naturally some waste products have to be sent back into the medium in return. This metabolic energy is then used for the synthesis of all this great variety of chemical components.

In this way the cell grows until a time comes when it is necessary for it to divide into two, or it will become too big. This is an intricate process, the details of which we only partly understand, but the end result is that after it is over we have two cells where we had one before. Moreover, the division takes place in such a way that in its essential features each daughter cell is identical, or almost identical, with the mother cell from which it came.

How is this done? The trick used by nature is to store the genetic instructions on a polymer—an immensely long molecule, whose total length in *E. coli* is about one millimeter, which is almost a thousand times the actual diameter of the cell. This long molecule is then copied exactly so that there are two daughter molecules in place of the single original mother molecule. Each of the two daughter cells at division gets one of these long molecules.

The chemical nature of these genetic molecules is very different from that of protein; they are in fact made of nucleic acid, in this case of DNA. The general plan on which nucleic acid is constructed is very simple. It consists of a very long backbone made of alternate groups of phosphate and sugar so that the backbone goes "phosphate — sugar — phosphate—sugar—phosphate—sugar" for very many units. To each sugar molecule is attached another kind of molecule, of which there are four different types. As we go along a molecule of nucleic acid we come across a sequence of these four side groups, which I shall call *A, a, B,* and *b*. We believe it is the precise sequence of these four letters which spells out the genetic information. In point of fact, in most cells nucleic acid consists of a pair of such chains wound round one another and linked together in such a way that each is the complement of the other. That is to say, where we have a big side-group on one chain, we have a small one on the other chain. Moreover the specificity of the fit is such that *A* can join on only to *a* and *B* can join on only to *b*.

The great advantage of this structure is that it enables a simple copying process to take place. In order to copy a piece of nucleic acid it is obviously necessary to have a supply of the four components which are going to be polymerized into the new chain. The real problem, however, is to get these components into the right order, so that the new message is an exact copy of the old one. The technique is quite simple. The two chains are first separated, and then each one acts as a template to guide the formation of a new companion chain. If at a certain point the old chain has a capital *A*, then a little *a* is inserted in the opposite place in the new

38

chain. If on the other hand it has a little *b*, then its companion will be big *B*. By this means one enzyme can direct the synthesis of any sequence of the four letters, being guided by the complementary sequence existing in the other chain. We thus start off with a pair of complementary chains; these separate and each one makes its own complement, so that we end up with two pairs of complementary chains, where we only had one before. Moreover the precise sequence of bases in these two daughter molecules is exactly the same as in the original mother one. The mechanism could hardly be simpler.

A point to notice is that one enzyme will do the entire job. It will copy any sequence whatever, provided that it is expressed in the four standard letters. As a result of the brilliant work of Arthur Kornberg and his colleagues, this mechanism, or something very like it, has been shown to operate in the test tube, using the enzyme purified from *E. coli* and a supply of the appropriate raw materials. There are still a number of things about the process we do not understand, not the least of which is the fact that the two chains are not lying side by side, but are wound round one another, and that in order for the replication to take place they must be at some stage unwound. Exactly how this happens we do not yet know. In addition, the process appears to be one of great precision. It looks as if there may be repair mechanisms that can correct mistakes made by the enzyme in the copying process, and others that repair damage accumulated by the DNA during its life in the cell.

On this picture, then, a gene is simply one stretch of the enormously long molecule of nucleic acid. The information making up this gene is written in the four-letter code, being the precise sequence of the four kinds of base along that particular bit of nucleic acid.

But clearly it is not much use having information unless it is used for something, so we must next ask, "What is it that genes do?" The answer is simple: the main function of genes is to direct the synthesis of protein molecules, each gene being used to direct the synthesis of a particular protein. However, the gene does not control this process directly. Instead a series of working copies of the gene is first made in a related material, also nucleic acid, but this time known as RNA. This enables the cell to produce copies of such genes as it needs at a particular time, rather than having to copy the information from all genes simultaneously. Each short length of RNA is then used as a message to direct the synthesis of a particular protein. The flow of information thus goes: $DNA \rightarrow RNA \rightarrow$ protein.

Before we can say how this is done we must first consider how proteins are made. Interestingly enough, the same general plan that is used in the construction of nucleic acid is used for making proteins. Protein molecules, too, consist of long chains, the backbone of which has a regular repeat and to which are attached, at regular intervals, various side chains. Chemically, of course, the backbone and the side chains are quite different from those of nucleic acid, and, in addition, instead of there being just four different types of side groups, in the case of proteins there are twenty. Proteins are made in fact by putting together the little monomer units, containing the side chain and the necessary link of the backbone, which are known as amino acids. We, ourselves, for example, can manufacture in our cells about half of the twenty kinds of amino acids we need. The rest of them have to be supplied to us in our diet, mainly in the form of the proteins of other organisms. These we break down into their constituent amino acids, which we then build up again to make our own proteins.

A typical protein usually contains a single chain, perhaps two, or three, or even four hundred or so amino acids long, but all made from the standard set of twenty units. The structure, and therefore the function, of the protein is determined by the exact sequence of these side chains—the amino acid sequence, as it is called—which is characteristic of each particular protein. The sequence dictates how the protein molecule folds up. This it does mainly by tucking inside all the side groups that do not like to be next to water, leaving outside all those that do. The process produces a complicated and rather definite structure for each protein, with various cavities and cunningly arranged active groups in just the right places so that the protein can carry on just that particular catalytic activity and no other.

It is possible in this way, following this rather simple general plan of a uniform backbone and a relatively small number of different kinds of side groups, to build up intricate molecules of great power and versatility using a relatively simple and uniform mechanism. It is for this reason that proteins are so important in the life of the cell.

You can see immediately that the important thing in protein synthesis is to make sure that the

42

amino acids are joined head to tail *in the right order* for each protein. This is what the gene determines. It is a much more difficult process than just copying nucleic acid, because we have, in effect, to translate the information from a four-letter language into a twenty-letter language, and this is by no means easy. Quite elaborate chemical equipment has to be used to do it. This equipment takes the form of largish particles, about the size of virus particles, which may be thought of as the "reading heads" for protein synthesis, which go on to the end of a particular piece of messenger RNA. As they move along the tape they read off the sequence information on the RNA, so that the protein is synthesized with its amino acids in the correct sequence. Although we do not know all the details of this particular process, we certainly know enough to understand it in outline. In addition, quite complicated auxiliary apparatus is needed to bring in the amino acids into the right places, and with the right specificity, in order to make the assembly process highly efficient.

Whatever the mechanism, you will see that it is necessary to have something like the Morse code to translate from the four-letter language of a nucleic acid into the twenty-letter language of the protein.

This code has recently been discovered, if not completely at least very largely, and is set out in Table 1. The code is in fact a very simple one. The nucleic acid message is read three bases at a time, so that the code is a triplet code. There are sixty-four such possible triplets, so that in most cases each amino acid corresponds to more than one of the triplets. There also appear to be special triplets for starting the chain and for stopping it, although these are at the moment only partly understood.

The over-all organization of the cell is thus very easy to grasp. The genetic message is carried on a polymer—on a molecule of nucleic acid—as the sequence of the four bases. Stretches of this message are then copied onto another sort of nucleic acid which acts as a tape to direct the synthesis of protein. Protein is made from twenty kinds of subunits, and the genetic message makes sure that these are joined in the right order for any particular protein. When the proteins are made they then act as catalysts (or sometimes as structural components of the cell) and control the many chemical reactions the cell needs for growth and division.

That is the mechanism in outline, but it is obvious that there must be fairly elaborate control

TABLE 1
THE GENETIC CODE

2nd letter

		a	b	A	B	
1st letter	**a**	14	16	19	5	a
		14	16	19	5	b
		11	16	21	?	A
		11	16	21	18	B
	b	11	15	9	2	a
		11	15	9	2	b
		11	15	7	2	A
		11	15	7	2	B
	A	10	17	3	16	a
		10	17	3	16	b
		?	17	12	2	A
		13	17	12	2	B
	B	20	1	4	8	a
		20	1	4	8	b
		20	1	6	8	A
		20	1	6	8	B

3rd letter

Each amino acid is coded by a triplet of three bases, as shown in this table, which is a compact way of setting out the sixty-four possible triplets.

The four bases of messenger RNA are denoted by the letters a, b, A, and B, and are identified:

a = Uracil   A = Adenine
b = Cytosine   B = Guanine

The twenty amino acids are identified by the numbers 1 to 20, with 21 standing for chain termination. The numbers stand for:

1. Alanine   8. Glycine   15. Proline
2. Arginine   9. Histidine   16. Serine
3. Asparagine   10. Isoleucine   17. Threonine
4. Aspartic acid   11. Leucine   18. Tryptophan
5. Cysteine   12. Lysine   19. Tyrosine
6. Glutamic acid   13. Methionine   20. Valine
7. Glutamine   14. Phenylalanine   21. End chain

For example the triplet bAB stands for No. 7. This implies that cytosine - adenine - guanine codes glutamine.

mechanisms to make sure that the various chemical reactions interlock properly. These control mechanisms are of two types: the first type makes sure that there is no excessive synthesis, or lack of synthesis, of the small molecules. In other words, the first mechanism does not affect the way in which proteins are made, but it does affect the rate at which particular proteins act. For example, a whole series of enzymes may be needed to produce the amino acid histidine. If ready-made histidine is now supplied to the cell from outside, then, if these enzymes go on acting at full speed, too much histidine will be produced. It is arranged, however, that when there is too great a concentration of histidine inside the cell the first enzyme of the chain is inhibited, and the production of histidine is slowed down. By this simple method, therefore, the cell arranges that there is never too little histidine and also never too much.

An ingenious feature of these control mechanisms is that they do not act on all the enzymes in a particular chain of enzymes, but usually act only on the first one, since if the action of the first one is slowed down the action of all the others is similarly slowed, because there is then only a small amount of substrate provided for each enzyme by the pre-

46

vious one in the chain. If there is a branch point in the metabolic sequence, such that one particular small molecule can be used for either of two purposes, then the control mechanisms usually occur just after this branch point, so that each of the two pathways can be regulated independently.

This mechanism does not affect the rate of *production* of the various protein molecules making up the enzymatic chain. This is done by a second mechanism, which is concerned not with the action of the enzymes but with their rate of synthesis. Again, we may take the example of histidine. If histidine is added to the medium, this has the effect of switching off the action of the genes that are responsible for making the various enzymes in the histidine chain. The control mechanism is such that the enzymes are switched as a group, so that as time goes on no more histidine-making enzymes are produced, and the enzyme molecules that remain are slowly diluted out as the cells grow and increase in number. If, at any moment, histidine is removed, then the synthesis of all these enzymes is switched on again, and after a comparatively short delay the cell will contain sufficient amounts of these enzymes to make all the histidine it needs from the raw materials supplied by the medium.

47

It is a rather general feature of this sort of control mechanism that it usually works not on a single enzyme, but on a group of them. In some organisms it is found that the genes for such a group of enzymes all lie together on the genetic material. Exactly how this control mechanism works we do not yet know, but we suspect that the fact that the genes lie side by side makes it probable that in such cases the control acts at the gene level and influences the amount of messenger RNA produced by such a group of genes.

In addition to these particular controls there must be general controls over the total rate of nucleic acid synthesis, the total rate of protein synthesis, and also over such matters as the growth of the cell and the division of the cell into two daughter cells. We are only just beginning to understand the details of these processes, but, as far as we can see, it should be easily possible to explain them in terms of the interaction of protein molecules and their products.

At this point it is convenient to go back and consider the nature of very simple viruses. Viruses are of many different shapes and sizes, but I shall consider here only the small ones, and these are usually either rods or spheres. Moreover, the plan

48

on which they are constructed is an extremely simple one. Inside each virus there is a long chain of nucleic acid which carries the genetic information. In most small viruses this is single-stranded RNA, although we know viruses which have double-stranded RNA. In addition some virsuses have DNA, the larger ones often being double-stranded and the smaller ones single-stranded. The exact nature of the nucleic acid is, therefore, not particularly important. The important point is that the genetic material is always nucleic acid of one sort or another. A typical size, for a small virus, is usually about six thousand bases long; that is to say, the genetic message, written in the four-letter language, is usually about six thousand symbols long.

This rather delicate genetic molecule is protected by a container of some sort which is made, in the simplest viruses, entirely of protein molecules. These make up the tubular wall of the rodlike viruses, or the spherical shell of the spherical viruses. In each case the wall is built up not of a single protein molecule but by using very many identical small protein molecules packed together in a rather symmetrical way to make a complete shell. Some of the genetic information in the virus is used to make this particular protein, though it

49

has other genes that code for several other proteins which are needed for the life of the virus, but are not part of its structure.

By one method or another the virus, having come in contact with the outside of its host cell, penetrates the cell, or at any rate allows its nucleic acid to penetrate, since for further infection the protein coat is usually no longer required. The virus then proceeds to take over, either completely or in part, the synthetic machinery of the host cell. The virus itself contains none of the elaborate apparatus for protein synthesis, which translates from the four-letter language of nucleic acid into the twenty-letter language of proteins. What the virus does is to use the chemical machinery of the host cell. By supplying it with the instructions contained in the viral nucleic acid it forces the cell to produce viral proteins which are needed both to disrupt the life of the cell and eventually to make finished virus particles, which can then come out of the cell and proceed to infect other cells.

We see, therefore, that the virus is the simplest possible living thing. It can multiply only in an environment that contains all the mechanism for protein synthesis, and this will not normally happen outside a living cell. Consequently to call a virus a living thing is really to stretch the term to the limit. It has some of the properties of a living system. It can multply geometrically and undergo mutations, and can, in this highly complex environment, exert a very specific influence, since it can direct the production of viral proteins which can then act in such a way as to influence the surroundings of the virus to its own advantage. It is thus possible for it to evolve by natural selection. On the other hand it is not possible for it to multiply unless it can find an environment where protein synthesis can go on. For this reason it is perhaps better not to call the virus alone a living system, but to consider the combination of the virus and its host as having the essential features we would normally look for. This is particularly so as the virus cannot metabolize in any way by itself and therefore has to rely on its host for supplying the energy necessary for the synthesis of more virus. This energy has to come from the metabolic processes going on in the host cell, and ultimately, of course, from the energy contained in the raw materials in the medium in which the host cell is growing.

At this point it is worth returning to our theme— the nature of vitalism—and asking ourselves which of the various processes we have been examining are likely to be the seat of the "vital principle" or of whatever vital process we think may be taking place.

It can hardly be the action of enzymes, because we can easily make this happen in the test tube. Moreover most enzymes act on rather simple organic molecules which we can easily synthesize. It is true that at the moment nobody has synthesized an actual enzyme chemically, but we can see no difficulty in doing this in principle, and in fact I would predict quite confidently that it will be done within the next five or ten years. The action of an enzyme, therefore, does not seem to present any special difficulty, although we always have to remember that the great specificity and subtlety shown by any particular enzyme is the result of a long process of natural selection. We could only synthesize a good enzyme, at this stage in our knowledge, by precisely imitating what nature has produced over the course of evolution rather than by designing one ourselves from first principles.

There does not appear to be an additional difficulty in processes that involve a chain of enzymes. Again, by working with enzyme mixtures we can get many of these processes to take place in the test tube. Just as we can, in principle, synthesize a single enzyme, so eventually it should be possible for us to synthesize every enzyme in a particular chain. Some of the enzyme systems (which I have not described in detail) depend upon the enzymes being part of particular structures; for example, the enzymes that provide metabolic energy for the cell are mainly housed in little organelles, in the cells of higher organisms, called mitochondria. These mitochondria have elaborate membranes and a certain amount of internal structure. It may be some time before we could easily synthesize such an object, but eventually we feel there should be no gross difficulty in putting a mitochondrion together from its component parts. This reservation aside, it looks as if any system of enzymes could also be made to act without invoking any special principles, or without involving material that we could not synthesize in the laboratory.

The same remarks apply to the synthesis of nucleic acid. At the moment it is not very easy, for technical chemical reasons, to synthesize a piece of nucleic acid with a defined sequence of bases, but it is already possible to make sequences of three or four or five bases and also repeating se-

quences—ones that go ABABABABABABAB . . .
and so forth. Again, it appears only a matter of
time before we can synthesize sequences of reasonable length.

This brings up a point which we shall find
recurs again. Whereas in principle we can conceive synthesizing a defined sequence of, say, six
thousand nucleotides to make up a piece of virus
nucleic acid, in practice the yields are likely to be
rather low. Even if we get a 99-per-cent yield at
every step, by the time we have carried out six
thousand successive steps the amount of finished
material will be very small indeed compared with
our starting material. We shall often find for biological objects that it is theoretically possible to
synthesize a particular molecule, but because of its
very large size it may be almost impossible to do
this in practice without taking an enormous
amount of time and huge quantities of raw material. Nature is able to make them because nature
has been at the job for so long, the process of natural selection having gone on for several thousand
million years. This has refined the structure of the
virus in a way that is not easy for us to do in the
much shorter time we have available in our labora-
tories. Although in principle it is possible to syn-

thesize in the test tube any particular piece of genetic material from its components, the longer it is, the more difficult it is, and the more remote the likelihood that it will be done in the near future.

So much for the structure of nucleic acid. The replication of nucleic acid can already be done in the test tube, although at the moment it is necessary to use the enzyme extracted from living cells as a catalyst. Here again, in principle, it should be possible to synthesize such an enzyme from the amino acids of which it is made, and this does not appear to be insuperably difficult. At the moment the actual synthesis is not particularly accurate. It may be that we have damaged the enzyme somewhat in getting it out of the cell, or possibly may even have the wrong one and be using what is really a repair enzyme. However, the recent work of Sol Spiegelman and his colleagues, using a little RNA virus, has shown that it is possible, in the test tube, to increase the amount of *infective* nucleic acid by a very large amount. This proves that true copying is taking place in the test tube. There is nothing, therefore, in the basic copying process, as far as we can see, which is different from our experience of physics and chemistry except, of

55

course, that it is exceptionally well designed and rather more complicated.

This leaves us, then, with very little that we have been unable to explain. It is true that we do not yet know exactly how membranes are made, or much about the metabolic pumps that pump molecules from the outside of the cell to the inside. We are also not very clear about the details of cell division. We only partly understand the assembly of the more complicated types of virus particle. But here again we hope to be able before long to make these processes happen in the test tube. The question of cell division also involves control mechanism, and, once again, we do not know precisely how control mechanisms act in biochemical terms. This is the problem that is being most actively studied at the moment, and we confidently expect that within the next ten years we not only will understand how these controls work, but will be able to reproduce them outside the cell.

It is difficult, then, having analyzed the cell into its various operations, to see which of the operations is in fact as mysterious as it appeared before we had all this detailed knowledge. What emerges quite clearly is that the cell is a very complicated object, and that it will indeed be extremely difficult for us to synthesize it from scratch. It will not be easy for us, therefore, to create life in the strict sense. On the other hand when we examine any bit of the mechanism, and see how it is made and how it works, there seems to be no difficulty in principle in synthesizing it ourselves, starting from rather simple chemical materials.

It looks, therefore, as if the borderline between the living and the nonliving does not cause us very serious trouble in explaining what we observe in terms of physics and chemistry. Naturally we cannot take over to higher organisms everything we have learned from bacterial cells and viruses, but it is a remarkable point, and one which should be stressed, that much of what I have described here does apply to higher organisms as well. In higher organisms we are reasonably certain that the main, if not the only, genetic material in the cell—at least using genetic material in the narrow sense of the word—is nucleic acid of one form or another. Right throughout nature we find that DNA and RNA are used in very similar ways to the way they are used in micro-organisms. DNA is made of the same four bases that are found in micro-organisms, and the same is true of RNA, so that the language that is used in the nucleic acid polymers is universal.

57

Even when small modifications are found to the bases these do not affect the way that they pair together, and it appears true that the pairing mechanism which is used in the replication of nucleic acid is identical in micro-organisms, plants, and animals, ourselves included. In exactly the same way, when we come to consider protein we find that, with only minor reservations, the same twenty amino acids are used throughout nature.

Of course the amount of genetic information needed to determine a human being is much larger than that required for a little bacterial cell. The DNA of *E. coli* is about 1 millimeter long—quite a good length considering that the cell is only one-thousandth of 1 millimeter in diameter. Each one of our own cells has far more than this, distributed among our numerous chromosomes. If all the DNA molecules from one of our cells were laid end to end, they would cover about two meters.

It is perhaps more striking to calculate how far the DNA would reach if we took the DNA molecules from *every* cell in our bodies. The distance is enormous, being comparable to the diameter of the solar system. It may seem remarkable that a small part of our cells can be spread out so far, but this is because on the scale of everyday life the

DNA molecule is so thin, being only about ten atoms across. It is really very self-centered of us to say that it is so thin. From the molecular point of view we are so fat compared with atoms. Put another way, in order to make an organism that can think about itself a vast number of atoms is needed.

There are two other striking ways to consider the DNA in our bodies. One is to calculate how much could be "written" using the DNA molecules from just one sperm cell. This comes to about five hundred large books, all different—a fair-sized private library. This is not surprising when we realize how complicated human beings are and how much information must be needed to construct one. The other calculation concerns the genetic material of the entire human race. If we took the DNA from just one cell of all the people alive today, how much space would it occupy? Of course, a lot of this would be repetitive since many of your genes are the same as mine. Nevertheless, it could all be contained in a volume the size of a rather large drop of water.

There is one feature of both nucleic acid and protein, and indeed of many biological molecules, that I have not yet mentioned. This is the fact that they are all of one "hand" and not a mixture of both. Organic molecules of a certain type can exist in two forms: one we may arbitrarily call right-handed; the other is the mirror image of the first one and can be called left-handed. It has been well known for many years that for any particular molecule only one hand occurs in nature. For example the amino acids one finds in proteins are always what are called the "L" or "levo" amino acids, and never the "D" or "dextro" amino acids. Only one of the two mirror possibilities occurs in proteins, and this is true whether the amino acids come from the proteins of a micro-organism, of a plant, or of an animal.

Finally, although we are not quite sure that the genetic code, which links together the four-letter language of nucleic acid and the twenty-letter language of proteins, is absolutely the same throughout nature, we know already that it is certainly very similar. These facts can only point to a common origin for all living things. Otherwise there seems no reason why there should not be differences in the systems as we find them today. When we come to consider how living things arose, this will be one of the important points we shall need to explain.

The Prospect Before Us

"That the fundamental aspects of heredity should have turned out to be so extraordinarily simple supports us in the hope that nature may, after all, be entirely approachable. Her much-advertised inscrutability has once more been found to be an illusion due to our ignorance. This is encouraging, for, if the world in which we live were as complicated as some of our friends would have us believe we might well despair that biology could ever become an exact science."

THOMAS HUNT MORGAN, 1919

WE HAVE SEEN THAT MOLECULAR BIOLOGY CAN already explain in outline many of the fundamental processes taking place in the smallest living things, so it is not unreasonable to ask what further knowledge we need in order to make quite sure that there is nothing unusual at this borderline between the living and the nonliving. One project that is being talked about at the moment could be described as the complete solution of *E. coli:* to find out everything about this one micro-organism, so that, whatever function we are studying, we can explain it in terms of the interaction of component parts whose structures we know.

It is clear that this is a major project. There are probably two or three thousand different genes in *E. coli,* and we shall need to know what each one of them does, and something about the molecular structure of the products of many of them. In

addition we want to understand all the control
mechanisms—how they work and what it is that
sets them off—the structure and function of the cell
membrane and the cell wall, and also the mechanism of cell division. This particular problem
will keep very many scientists busy for a long time
to come. Nevertheless, it is the sort of project on
which large numbers could usefully be employed,
because it has a fairly well-defined goal, and there
is no reason why substantial knowledge should not
accumulate if sufficient people were to work on it.

It seems to me far more important to be able to
understand a living cell in this way rather than to
worry about whether we could synthesize it completely, starting from the elements. It is almost
certain that by the time we have all this knowledge
we shall be able to see that we could, if we wished,
synthesize any of the bits. The problem of synthesizing it all and then putting it together on such
a small scale—although a fascinating one—seems
hardly worthwhile when one considers the immense amount of labor that would have to be
involved in order to make everything synthetically.
It would be more reasonable to see if we could take
it apart and then put the pieces together again,
using components from broken cells, together per-

haps with a few that had been made by chemical synthesis.

Even if we had all this knowledge this would by no means tell us about the behavior of organisms having many cells, and especially organisms like ourselves. The subject of morphogenesis, the origin of the structure of higher organisms — how our bones and nerves and muscles, for example, occur in the right places, how our eyes are formed—this subject is again one that is being actively studied and about which knowledge is likely to increase. We shall undoubtedly find that there are additional control mechanisms in higher organisms which do not occur in micro-organisms. We already know something of these mechanisms; for example, the nature of hormones and a little, though not very much, about how they act. In addition we shall have to learn how one cell recognizes another cell, and also how cells migrate so that they can get into the right place, not only in the growing animal but also during the various repair processes that come into action when an animal is wounded.

All this will take a very considerable time, and it is clear that we shall make many interesting discoveries on the way. However, it is difficult to foresee anything which, with our present knowledge, would be impossible for us to explain. This springs from the fact that we know that the cell can make a great variety of proteins, and we also know that proteins are molecules of great versatility and subtlety. They are quite capable of carrying out, at the molecular level, the processes that are needed in order to give the cell its behavior on a larger scale. We already know something, for example, about muscular contraction and how the two sorts of protein filaments slide past one another to turn chemical energy into work. Even if we do not fully understand the details, we are confident that movement of this sort can easily be explained on a physical basis. The same is true for nerves. Again it does not seem that what we know will be difficult to understand in terms of the types of molecules which the cell has available to it.

The other great field of knowledge that will undoubtedly expand and that is associated with molecular biology concerns the origin of life. However we define the point of origin, it was a step that took place an exceedingly long time ago, probably about three or four thousand million years ago. We have a general picture of what the world was like at that time. In particular we believe that the atmosphere did not contain oxygen and was reducing in nature, and that under the action of electrical discharges and of ultraviolet light the gases of the atmosphere were converted into simple organic compounds. These then dissolved in the sea so that the primitive oceans of that period can be thought of as a rather dilute soup, consisting of many different rather simple organic molecules.

One might wonder why the soup did not go bad, but the answer is quite obvious. There was no living thing to make it go bad! It is true that most of these chemical molecules are, on a long-time scale, somewhat unstable and would break down by themselves. Presumably some sort of steady state was reached between the synthesis of molecules by electric discharges and similar processes, and their spontaneous breakdown to other compounds.

Eventually these simple molecules must have become joined together into polymers, and these polymers must have begun to interact in interesting ways. The crucial point is the one at which natural selection could begin to act, since from then on the system could go on improving itself. For example, as it used up the supply of complicated organic molecules in the ocean, it could improvise the enzymes necessary to make these compounds for itself from simpler chemical molecules

67

still available in the ocean. *Before* natural selection could operate everything would have to occur more or less by chance, and even though we have vast quantities of space available to us and great stretches of time, there is a limit to the number of things we can imagine to have happened by accident. The actual "origin" was probably a rather rare event; indeed as I have already argued it probably took place only once.

The present state of this subject is moderately encouraging. Chemists can already explain how many of the simpler organic molecules that would be needed—the amino acids, the sugars, the bases, and so forth—could have come into existence by a variety of reasonably plausible mechanisms. It seems likely that before long we shall be able to explain the origin of the simpler molecules making up the soup. It is not yet clear how the slightly more complicated activated forms of these molecules were made, but this again is a problem that chemists will be considering in the next few years, and we may hope that a number of interesting possibilities will be discovered by doing model experiments in the laboratory.

Once such activated precursors were formed they

could become joined together by polymerization processes. The concentration of these organic molecules, however, would appear to have been rather low, and some thought will have to be given to the mechanisms whereby the relatively thin, weak soup could be concentrated, in special places, into a much richer and more nutritious form of broth. It is the next steps that seem difficult, and that may not be easy to study. At the moment when natural selection started, was there only nucleic acid and no protein? Or, on the other hand, was there only protein and no nucleic acid? The difficulty of these alternatives is that if we had protein alone it is not easy to think of a simple replication process, whereas if we had nucleic acid alone (which would make the replication easy) it is difficult to see how nucleic acid could provide the necessary catalytic activity. A third possibility, which to my mind is rather promising, is that when natural selection began both nucleic acid and protein existed, and that the synthesis of protein was crudely coupled to nucleic acid in the same sort of way as it is today. At first sight it seems highly unlikely that this complicated mechanism could have arisen by chance, but it is really quite possible that some primitive

69

version of it started in that way and although not perfect was sufficiently accurate to enable the system to get going.

The real difficulty about the origin of life is that the experimental evidence showing what happened has long ago disappeared. All we are left with is a certain amout of frozen history in the organisms as we see them today. This is going to make it scientifically very difficult, because it is inevitable that there will be more theories than there are facts to disprove them. I can foresee that the subject, instead of developing in a proper scientific way, will become almost theological. Various schools will arise, with different theories, each of which will be vigorously supported, but there may well be insufficient facts to enable us to choose between them. Scientists working in this area will have to show a certain amount of restraint in putting forward ideas, and rather more effort than usual in looking for evidence, if the subject is not to get hopelessly bogged down in a morass of unproved theories.

There is still the hope that there may be direct experimental evidence somewhere, if not hidden in some very cold rocks in parts of the earth then possibly in meteorites, or perhaps on the surface of the moon or of Mars. It used to be said, in the nineteenth century, that we would never know of what the stars are made, since we could never visit them. We all can see how foolish a prophecy this was. One can only hope that similar statements about the difficulties of obtaining evidence of the origin of life will be shown, in the not too distant future, to be equally pessimistic.

We must now consider how much is known about that other sensitive area, the higher nervous system. The working of our brains is based on the working of the nerves that make up our brains and the way they influence one another. Many of the properties of individual nerves are tolerably well understood. Perhaps the most important property about the signal sent down a long nerve is that it has an "all-or-none" character. It is a self-regenerating signal, which once it has started at one end of the axon goes along with a constant velocity to the other end, and with a constant amplitude, so that no information is contained in the actual amplitude of the signal. The information sent along the nerve is conveyed solely by the frequency of these pulses. This is probably the most important single discovery since it was found that the brain was actually constructed out of in-

71

dividual nerves and not on some other principle.

We also know that signals can pass from one nerve cell to another, at points known as synapses, and something of the mechanism of the transmission at the synapse. This can sometimes be electrical, but it is more often chemical. It may be either a signal of activation or one of inhibition. Knowledge of the chemical nature of these signals, and of the way they act, is accumulating rapidly; but there are many delicate features about the interaction between two nerve cells which are certainly only very dimly understood.

Nor do we know very much of the general plan of the nervous system. We can see how great tracts of nerve fibers go from one place to another, but it is clear that such an intricate mechanism must be based on some general principles of organization. In our brains the nervous system seems to be organized into rather thickish sheets. It is as if it were a structure based on two and a half dimensions rather than on either two or three, but apart from one or two hints from recent work the way in which nerves are grouped together is not really understood. Nor is it understood how nerves know how to find their way to the right place and make the right contacts, either when they are growing in the embryo or when, in some organisms, they are regenerating. This information must largely be carried in the genetic material, and we do not have the remotest idea, at the moment, of how this genetic information is expressed. It seems unlikely that there can be a separate gene for every contact between a pair of nerve cells, as there would not be enough genes to go round. Presumably it is done on some repetitive principle, with general instructions and then more special instructions, but the details of this are at the moment completely unknown.

When we come to consider how the brain learns—how it remembers—we are even more in the dark. Research is going on at the moment to try to find something of the chemistry involved in this process, but it would appear to me that this is going to be a long and difficult task. I rather suspect the problem will not be finally solved until we know the nature of the hereditary mechanism that constructs the nervous system, as memory may perhaps be a small modification of the system laid down by the genes and thus understandable only when we understand how these genes act.

It is clear from what I have said that our knowledge of the nervous system is at the moment in an

73

exceedingly primitive state. It is not quite fair to make the comparison with molecular biology, which has made rather rapid progress in the last ten or fifteen years. This is because the things we most want to know are rather different in the two cases. In molecular biology we were mainly interested in the borderline between the living and the nonliving, and we were looking for rather simple explanations of rather simple processes. We have not yet worried about complicated operations of the sort we find in morphogenesis, like the making of the fingers of the hand, or the growing of hairs on the head. When we come to the nervous system we already understand, in a general way, the workings of the basic elements, the neurons. In this case we want to understand *ourselves*—how our brains work and why we are conscious. Thus we need to grasp the behavior of the organism as a whole, and all the complex interactions that go into making that behavior. It is for this reason that I think the study of the nervous system is going to have to continue for a considerable period before it can answer those questions about ourselves which most interest us.

One other approach to the nervous system, which is progressing very rapidly today, is the

74

study of man-made computers. The first point to make here is that, however we compare a computer with the brain, we find that even the most complex of today's computers are far, far simpler than the human brain. This can easily be shown by counting up the number of elements in the two. It is found that there are about a thousand times more nerve cells in the brain than there are elements in even our largest computers. Again, the brain has an enormous advantage in that it can be packed into a small space and work on very little power. However, modern computers are getting rather smaller in physical size because of the increasing use of transistors and of printed circuits. Although it will not be possible in the near future to make elements as small as neurons, at least the difference in size is not as apparent as it was a few years ago.

It is most important to realize that modern computers are based on quite a different system of operation from anything we can see in the brain. Such computers work on a binary system and are extremely accurate. Our brains, on the other hand, show no signs that we can see of working on a binary system, and in addition are very inaccurate. However the brain is able to work with the loss of quite a fair number of its nerve cells (every day some die as we grow older) and is presumably inaccurate because it is made so that the loss of one particular element does not upset the functioning of the brain as a whole. Computers, on the other hand, are usually designed in such a way that, as far as possible, all their elements must be functioning satisfactorily.

The great advantage a computer has over the brain is that its basic rate of working is very much quicker. Its "pulse rate" is approximately a thousand times faster than the corresponding rate of sending signals in the brain. Consequently even a rather small computer can undertake a task that the brain cannot tackle in any reasonable time. On the other hand the brain is a multiple-input device. Over a million nerve fibers go from the eye to the brain, and many of these can be carrying signals at the same time. Thus the elements of the brain act to some extent in parallel, whereas much of the working of computers occurs in series. A computer is made in such a way that it can concentrate on doing one particular job at a time, or at the most a small number of different jobs. It is for these reasons that computers can do straight calculations more rapidly than we can with our brains, whereas our brains can do certain complex jobs, like the recognition of a friend's face, surprisingly quickly.

It is amusing to note the sort of things modern computers can do. One, for example, has been programed to play checkers. This is a type of program in which the machine evaluates each game after it has been played and then adjusts its strategy to optimize its method of playing. For a time it was even able to beat the man who wrote the program. Several people have written programs that allow a computer to play chess, which is a far more difficult game. So far such machines can only play very average chess. It seems likely that with more sophisticated programs a computer could be made to play quite a good game, and there are those who think that within the next twenty years or so a machine might become world champion.

A program has also been written to prove various theorems of Euclid, given some general instructions. One of the remarkable things that came out of such a program was that the machine invented a new proof which was simpler than the one that had been used for the last two thousand years. It was in fact so simple that nobody had noticed that you could prove this particular theorem in this way.

Programs are also actively being written for teaching purposes. One of these, for example, will give a young doctor a problem, instruct him what (imaginary) tests to carry out, cross-examine him on what he has done, and criticize his diagnosis, telling him sometimes to try again, or to go back and examine different sorts of evidence. A very amusing one has recently been written which simulates the behavior of a psychiatrist talking to a patient. The great advantage of this situation is that the "patient" is not supposed to ask the "psychiatrist" a question, whereas the "psychiatrist" can always initiate a new topic if he does not wish to continue along the lines of the previous one. This makes it very easy to write the program, since the computer can take cues from quite simple words. Here is an example of an exchange between a real person and an IBM 7090 computer. The program was written by Joseph Weizenbaum, visiting associate professor of electrical engineering at the Massachusetts Institute of Technology. It was designed to illustrate the technical characteristics of a new time-sharing system by which up to thirty persons can use the same computer at the

78

same time. The interview takes place by typewriter. The "patient" types out what he wishes
to say in lower-case letters, and then presses a
button twice, whereupon the machine replies, also
by typewriter, but in upper-case letters; and so the
conversation goes on.

In Which a Distraught Young Lady Consults the Machine*

men are all alike
IN WHAT WAY
they're always bugging us about something or other
CAN YOU THINK OF A SPECIFIC EXAMPLE
well, my boy friend made me come here
YOUR BOY FRIEND MADE YOU COME HERE
he noticed that i'm depressed much of the time
I AM SORRY TO HEAR YOU ARE DEPRESSED
he's right. i am unhappy
DO YOU THINK COMING HERE WILL HELP YOU NOT
   TO BE UNHAPPY
i need some help, that much is clear
WHAT WOULD IT MEAN TO YOU IF YOU GOT SOME
   HELP
perhaps i could learn to get along with my mother

*Quoted from the *Harvard Review,* III, No. 2, 70-71.

TELL ME MORE ABOUT YOUR FAMILY

my mother takes care of me

WHO ELSE IN YOUR FAMILY TAKES CARE OF YOU

my father

YOUR FATHER

you are like my father in some ways

WHAT RESEMBLANCE DO YOU SEE

you are not very aggressive but i don't think you
   want me to notice that

WHAT MAKES YOU THINK I AM NOT VERY AGGRESSIVE

you don't argue with me

WHY DO YOU THINK I DON'T ARGUE WITH YOU

you are afraid of me

DOES IT PLEASE YOU TO BELIEVE I AM AFRAID OF YOU

it pleases my father to think i am afraid of him

WHAT ELSE COMES TO MIND WHEN YOU THINK OF
   YOUR FATHER

bullies

DOES THAT HAVE ANYTHING TO DO WITH THE FACT
   THAT YOUR BOY FRIEND MADE YOU COME HERE

You can see that at one point the "patient" used
the word "like." This is probably a cue for the
machine to say, "WHAT RESEMBLANCE DO YOU SEE."
By use of such simple clues and other tricks it is
possible to make a dialogue that sounds reasonably

convincing, using a simple program. There is even talk that it may be possible to provide a service for disturbed people, so that they can have someone "sympathetic" to talk to. However, people who have spoken with such a computer find it an emotionally unnerving experience, partly because of the somewhat superior attitude of the machine, which is apt to ignore what one has said and simply ask a new question from time to time. Nevertheless, it is clear that with ingenuity it will be possible to extend this type of programing in quite a number of directions.

There is a story going around that one computer can now play a reasonable game of table tennis. This is done by hitching it up to a mechanical arm (developed for handling radioactive materials) and also to a couple of television cameras which can register the position of the ball at any particular time. It is said that as long as the other player does not put too much spin on the ball the computer does fairly well!

Nevertheless there are a number of simple operations which are still giving trouble to computers. One is the ability to read letters. It is not at all easy, apparently, to write a program such that a machine can recognize, say, the letter *s* in

81

various sizes, shapes, and forms. Even reading standard print seems to give trouble to the program writers. However, we can expect that this is the sort of difficulty which will be overcome in the next few years.

I think the really striking thing that will come from the use of computers will occur when they have reached a very high degree of sophistication and one can have a computer behind a screen with which one could carry on a conversation. The standard question is, "How will one find out that one is talking to a computer and not to a real person? Would the computer, if questioned, say that it was conscious and describe what it meant by this?" Of course the computer might have to be "educated," but then so do human beings. There are many people who believe that it will never be possible to simulate a human being in this way, but those who work on computers tend if anything to take the opposite point of view and think that we shall see this development in our lifetime.

Even if we do not, I am sure that interaction with computers will become a common part of everyday life. After all, we can easily get the computer to type out what it has said, and then get some actor to read it. It would be quite possible

82

to have a television program in which people spoke to the computer; their words would simply be typed down and passed to the computer by some intermediary; the machine would then reply on the typewriter, and an actor would read out what the machine had said. I prophesy that before long there will be a television program of this kind and that it will be a sensation, provided of course that the program is written with sufficient ingenuity. In fact I think writing programs of this sort will become a new literary occupation—we may have a program for, say, literary criticism which would certainly be great fun to devise. Alternatively someone might try and write a program for a seduction scene. Before long I can well imagine that people will put two machines together and see how they talk to each other. It would be very amusing to get the seduction program to talk to the psychiatrist program, and it might lead to some explosively funny situations!

But leaving aside all these fancier applications we have to realize, as has already been stressed by a number of people, that machines are going to take on many of the functions of human beings, and that it is going to be quite disturbing for us to associate with them. There are people, Fred

Hoyle for example, who believe that machines will eventually take control of our civilization; but even if that does not happen it could be argued that what will arise is a symbiosis between machines and men, in which the main function of the men will be to reproduce and to tend the machines. I doubt myself whether we shall reach quite such a stage, but nevertheless I am convinced that it is going to be quite upsetting having to associate with very complicated and sophisticated machines, and that this development is likely to happen during our own lifetime.

I think the people who are most disturbed by this sort of thing are those who believe in some sense in the soul. What is never very clear is what is meant by "the soul," but one of its attributes appears to be that it can associate with the body but is separate from it, and, in particular, that it can in some circumstances exist separately from the body and especially, so many people think, after death. One difficulty about the soul is to know when it originated in evolution—most people would agree that all human beings have souls (though no doubt there are a few eccentrics who think that they are denied to women), but it is not at all clear whether a chimpanzee or a dog can have one. It is noticeable that philosophers who keep dogs, and are fond of them, are more inclined to attribute souls to them than those who are not animal lovers. And if a dog, why not a worm, and so on? In other words, those properties which we associate with an organism which make us think it might have a soul do not spring fully fledged at some point in evolution, but appear to come gradually.

To take another familiar difficulty, does a baby have a soul? And if so, does it have it before birth, and at what moment does· it get it, since it seems hardly likely that the unfertilized egg has a soul in the sense of which we are talking. Of course there are standard religious answers to some of these questions, but they appear to me to be arbitrary nonsense.

Those who think the soul can survive after death are sometimes prone to believe in extra-sensory perception (ESP), and imagine that two minds can communicate directly with one another by some unknown mechanism, probably one with a nonphysical basis. I have even been approached by a well-meaning and enthusiastic modern clergy-man in Cambridge, who suggested that there must be some fascinating relationship between DNA

OF MOLECULES AND MEN

and ESP. But then, as far as I could make out, he thought that ectoplasm was good evidence in support of the Christian faith.

The most striking thing about the work of the last thirty years on ESP has been its complete failure to produce any technique whatsoever which is scientifically acceptable. There is no known way, by a special screening procedure, by the use of drugs, or by any other method, to discover people who can communicate in this way, and be proved to skeptical observers to do so. Not one truly reproducible experiment has been devised although the record is thick with fakes and sloppy experimentation. We must conclude either that the phenomenon does not exist, or that it is too difficult to study by present methods, or that the people who work on these problems are hopelessly third-rate. ESP has all the appearance of a completely "void" science, like astrology, in which no genuine experiments exist, and the only "results" are due to bad experimentation or to faking, either by subject or experimenter, conscious or unconscious. This background level, incidentally, occurs in all genuine subjects—there is about one fake in biochemistry every year or so—but in respectable sciences the "noise" is well below the

86

level of the "sense." In ESP the sense seems to be absent, and only the noise remains.

I myself, like many scientists, believe that the soul is imaginary and that what we call our minds is simply a way of talking about the functions of our brains. The real difficulty comes from the vividness of our experience of consciousness, and even that is a matter to some extent of degree, since we can be conscious to various extents, either when we are half awake or when we are sleep-walking. I also find it disturbing that we dream every night and retain so little of our dreams. Recent work has shown quite conclusively that everybody dreams each night for quite considerable periods and that most of these dreams are forgotten unless one is awakened while they are going on. It seems remarkable to me, in spite of everything Freud said many years ago, that we dream so much and yet remember so little about it.

An area of research that is likely to lead to interesting consequences comes from dividing the brain into two. This is an operation that can be performed quite easily on monkeys, and is occasionally done on human beings for one medical reason or another. It has the effect of producing a person who has to some extent two brains. The

87

peculiarities of such people spring mainly from the fact that only one side of the brain is concerned with speech, and consequently only one side can easily communicate with the outside world. Such people are not noticeably different in their behavior unless special tests are done, but if precautions are taken so that one side of the body cannot communicate with the other side, as by separating the two fields of vision, then it is possible to communicate with one half of the brain and not with the other. In this way Sperry and his collaborators have shown that both sides of the brain can respond and learn and behave in the way that one expects of a human being, except, as I have said, that the speech centers are concentrated on one side only. Curiously enough this has been taken by J. C. Eccles to imply that in some way the soul is indivisible, but it seems to me that (if it were ethically acceptable) one might try to train such a body to become two people. If, for long periods of time, one could prevent the two brains from communicating with one another, one could perhaps convince one brain that it was in the same body as another brain—in other words, one could make two people where there was only one before. Whether this can actually be done remains to be seen, but I think it would be almost as disturbing to us as if identical twins had been suddenly produced for the first time.

A constant theme of this book has been that of natural selection. It is not true, of course, that mankind is evolving at the moment only by natural selection, because ever since man was able to communicate and form societies another form of evolution, social evolution, has been taking place which is very much faster and in many ways more effective. Nevertheless, much in our nature has evolved under the pressure of natural selection alone, and these pressures still exist today. Natural selection being a slow process and civilization being fairly recent, it follows that a lot of our behavior was evolved in a period when human beings acted in a way rather differently from the way they do today. For example, much of our aggressiveness probably springs from behavior selected when man was living in small groups, probably in constant competition with each other. The same is probably true of much of our sexual behavior and explains many of the difficulties and dilemmas that we find in our marriage laws and in our sex laws in general.

I think it is difficult to overemphasize the im-

portance of teaching natural selection, both in
schools and in universities, so that every member
of our culture has a clear and firm grasp of the
principle involved. It is, I think, one of the
scandals of the United States that there are still
statutes on the law books of certain Southern
states which formally forbid the teaching of evolution in schools. Even though such a law may be
to some extent a dead letter, it is certainly most
reprehensible that it still exists. It is remarkable
to me that there is not more protest about the
situation.

Personally, I myself would go further, and
think it is also regrettable that there is so much
religious teaching. It is true that the situation in
the United States is nothing like as bad as it is in
Great Britain, where religious instruction, I am
sorry to say, is compulsory in all schools supported
by public money. Since much of this instruction,
from the point of view of most educated men, is
utter nonsense, it seems to me particularly dis-
tressing that this should be the one compulsory
subject of British education.

Nor unfortunately are many British universities
much better in this particular matter. It is true
that many of them have inherited a religious tra-

dition, but this does not easily explain the tre-
mendous institutional suport given to religion by
such a body as Cambridge University, and the col-
leges that form a part of it. The fact that many of
the senior members of the university believe that
what is taught and propagated in this way is really
beneath contempt intellectually is not apparently
enough to prevent the continuation of these an-
cient practices. Of course, if a university is founded
by a number of private people, there is no reason
why they should not propagate their religious beliefs if they wish to do so, but it is quite another
matter when a university is a public institution
and claims a wider body of support. The business
of the university has been well stated by Lord
Annan, the Provost of King's College, Cambridge,
as "intellect, intellect, intellect." Whatever other
functions the university may have, there is no
place for the support of half-truths and falsehoods,
and it is a remarkable thing that so many intel-
lectuals are hypocritical about this matter, and
either shrug their shoulders and say that they
are not personally involved, or feel that it is a
matter of no importance. The young, I am pleased
to note, take nowadays a healthier view of this
cynicism and feel that they can hardly respect their

91

elders while the latter tolerate such hypocrisy.

One should perhaps state clearly why questions like this are important. When our culture was firmly based on Christian beliefs and practices, answers were provided to many fundamental questions, and it was not thought necessary that science should do more than dot the *i*'s and cross the *t*'s for those answers. The position is quite different today. Today we know that everything we knew yesterday about questions of this type is almost certainly untrue. The intellectual should be concerned with questions such as "What are we?" "Why are we here?" "Why does the world work in this particular way?" It is remarkable to me that there is not more urgency to answer questions of this kind. I think this situation can only spring from the fact that most people believe either that these questions have been answered already, in some way or another, or alternatively that the answers are perhaps too difficult to understand. It would be a much healthier state of affairs if instead of the United States and Russia competing in an arms race they competed in a knowledge race; if it were to be regarded as a matter of national prestige that we could understand the nature of life, for example, rather than mounting enormous and costly space programs to go to Mars, although I would be the first to agree that going to Mars may perhaps help us a little way toward answering the question of the nature of life.

Once one has become adjusted to the idea that we are here because we have evolved from simple chemical compounds by a process of natural selection, it is remarkable how many of the problems of the modern world take on a completely new light. It is for this reason that it is important that science in general, and natural selection in particular, should become the basis on which we are to build the new culture. C. P. Snow was quite right when he said there were two cultures. (I do not wish to argue here whether there are two, or three, or four, but simply that there is more than one.) The mistake he made, in my view, was to underestimate the difference between them. The old, or literary culture, which was based originally on Christian values, is clearly dying, whereas the new culture, the scientific one, based on scientific values, is still in an early stage of development, although it is growing with great rapidity. It is not possible to see one's way clearly in the modern world unless one grasps this division between these two cultures and the fact that one is slowly dying and the other, although primitive, is bursting into life. University administrators should try to see that their universities become centers for the propagation of the new culture, and not merely homes for propping up an aging and dying one.

For this reason I believe that all university students should be taught a subject that might be called "The Map of Science." This would not only describe the broad nature of all the various sciences and the way they are related to each other (with a few selected illustrations from each to bring life to the description) but would also show how developed each science is and which areas are relatively understudied.

Such a course would clearly demonstrate that while, say, mechanics or optics are very well explored, much of biology is still almost virgin territory. It would encourage students to consider questions to which we do *not* yet know the answer, but which we think there is a hope of answering within the next three-score years. Now there are some questions that affect us far more personally than others, and among these the working of the brain certainly ranks high. It can be confidently stated that our present knowledge of the brain is so primitive—approximately at the stage of the four

humors in medicine or of bleeding in therapy
(What is psychoanalysis but mental bleeding?) —
that when we do have fuller knowledge our whole
picture of ourselves is bound to change radically.
Much that is now culturally acceptable will then
seem to be nonsense. People with training in the
arts still feel that in spite of the alterations made in
their lives by technology—by the internal combus-
tion engine, by penicillin, by the Bomb—modern
science has little to do with what concerns them
most deeply. As far as today's science is concerned
this is partly true, but tomorrow's science is going
to knock their culture right out from under them.

In addition to a general knowledge of science it
is desirable for everyone to have studied one particular branch of science in rather more depth.
This will automatically happen to science students,
but it is an interesting question which subject
should be taught to liberal arts students to give
them some insight into science and the scientific
method. Snow originally suggested that the best
educational touchstone was the Second Law of
Thermodynamics, but he was astonished to find
that his literary friends thought this term a joke.
It is also fairly difficult to understand without a
good grasp of mathematics. More recently he has

suggested the structure and replication of DNA. Any schoolchild can grasp this, so that it is particularly suitable to teach to literary people. In addition it is at the root of biology.

However, I am inclined to agree with J. Bronowski that a fascinating but neglected subject for general education is animal behavior. This is fairly easy to learn, as the concepts are not too strange for us. It has in recent years become a science—that is, hypotheses can be erected which can be disproved by experiment—and is in a period of growth that is likely to continue for many years to come, so that students would have the advantage of learning a living and developing science which they could follow after they had left the university. Moreover it teaches one of the great lessons of science in a rather personal form, namely that the familiar is not always what it seems. It is after all remarkable that so many people have for so many centuries kept cats and dogs and yet have learned very little about their behavior except some dubious "insights" based on traditional ideas about human behavior.

Animal behavior also demonstrates the importance of learning about ourselves by studying scientifically creatures lower in the biological scale.

For all these reasons it seems to me an excellent subject to teach to liberal arts students as part of their education during their first year at a university. And not only to liberal arts students. The cultural ignorance of chemists, for example, is sometimes almost beyond belief. There is every reason to give scientific experts a broad outlook in science which at present they so often lack.

And so, finally, we come to the question, "Is vitalism dead?" It seems to me that, reluctantly, we must answer, "No." While there are intelligent people alive who sincerely believe in vitalistic ideas, even though they are fully acquainted with the scientific knowledge on the subject, we must conclude that vitalism is still alive. There remains the question of how we can disprove it, assuming, as I believe, that all these vitalistic ideas are untrue and will eventually be shown to be so by increased scientific knowledge.

Let us look, then, in turn at our three sensitive areas. Turning first to molecular biology, it is clear that if we are to be quite certain that vitalistic concepts can play no role here we shall have to have more knowledge. But this, I think, will surely come because of the large number of people actively engaged at the present time in working on

97

these problems, and also because the knowledge we have already makes it highly unlikely that there is anything that cannot be explained by physics and chemistry. Our second difficult area—the origin of life—presents us with rather a different problem. In this case it may be difficult to get the knowledge we require in order to make sure that nothing unusual happened at the earliest stages of life on earth. It may be some time before this subject develops, especially as not very many people are working on these problems at the present time. However, I think that the knowledge we shall acquire of molecular biology will ensure that relatively few people will worry about the origin of life, since these two areas of study are so closely related.

It is quite a different matter when we come to the nervous system. Here vitalistic ideas not only are commonplace among educated laymen, but are held by several of the leading workers in this field. As I have stated, this is relatively speaking a scientifically backward area of study; many more workers are required and much more knowledge is needed before we shall be able even to clarify the major questions we need to ask about our brains, our behavior, and our strange feeling of being conscious. When this knowledge has been obtained, and when the study of fast computing machines has advanced even further (and this is likely to happen very rapidly) , then I think that vitalistic ideas about the brain will grow to look as peculiar as vitalistic ideas now seem in molecular biology. Exact knowledge is the enemy of vitalism.

Provided, then, that scientific study continues on a considerable scale, we can foresee a time when vitalism will not seriously be considered by educated men. Will vitalism then be dead? The answer I think is that it .will be dead, but its ghost will remain. It appears impossible to remove completely those beliefs to which the human mind is prone. A lunatic fringe always remains. There are still people today who believe that the earth is flat, in spite of all the enormous accumulation of scientific evidence to the contrary. And so to those of you who may be vitalists I would make this prophecy: what everyone believed yesterday, and you believe today, only cranks will believe tomorrow.

FRANCIS CRICK, distinguished British molecular biologist, won the Nobel Prize in Medicine in 1962, with J. D. Watson and M. H. F. Wilkins, for studies of the molecular structure of DNA, the nucleic acid within the living cell that transmits the hereditary pattern. He was also the recipient of the Lasker Award in 1960, with Watson and Wilkins, and the Gairdner Foundation Award, Toronto, in 1962.

Now a Laboratory Scientist at the Medical Research Council Unit for Molecular Biology in Cambridge, England, Dr. Crick is also a nonresident Fellow at the Salk Institute for Biological Studies in San Diego. He has been a visiting lecturer at the Rockefeller Institute in New York and has twice been a visiting professor at Harvard University, and has lectured extensively in this country and abroad. In 1959 he was elected a Fellow of the Royal Society.

Born in England in 1916, Dr. Crick earned his bachelor's degree at University College, London, in 1937, and his doctorate at Caius College, Cambridge University, in 1953. His education was interrupted by World War II, during which he served as a scientist with the British Admiralty.

Dr. Crick is the author of numerous papers and articles on molecular biology published and translated in many scientific journals.